I0468709

ZOO ANIMALS COLORING BOOK

Realistic Coloring Book for Adults, Animal Coloring Book for Adults containing 40 Advanced Coloring Pages

Realistic Animals Coloring Book: Vol 8

by Amanda Davenport

ISBN-13: 978-1530804542

ISBN-10: 153080454X

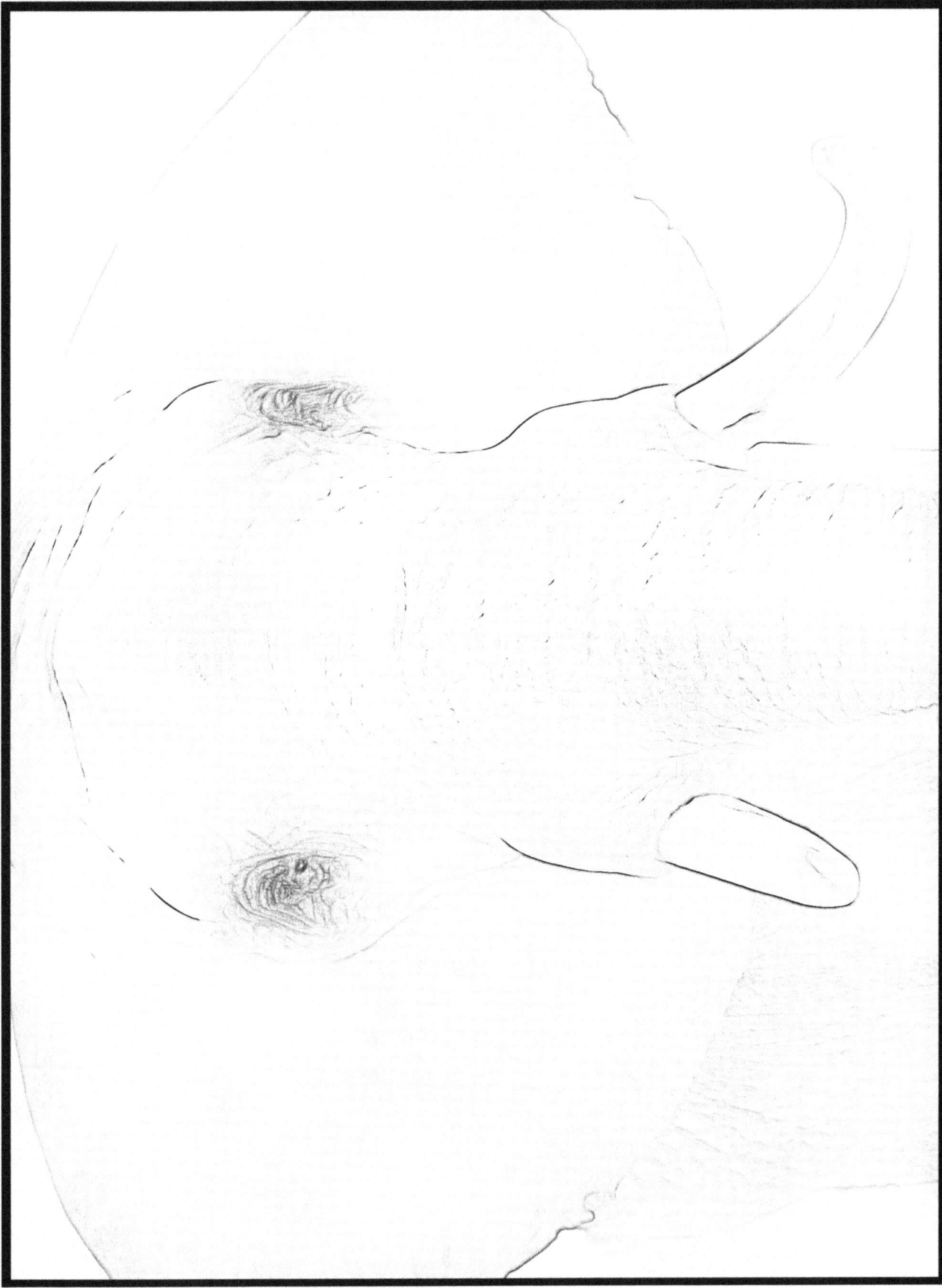

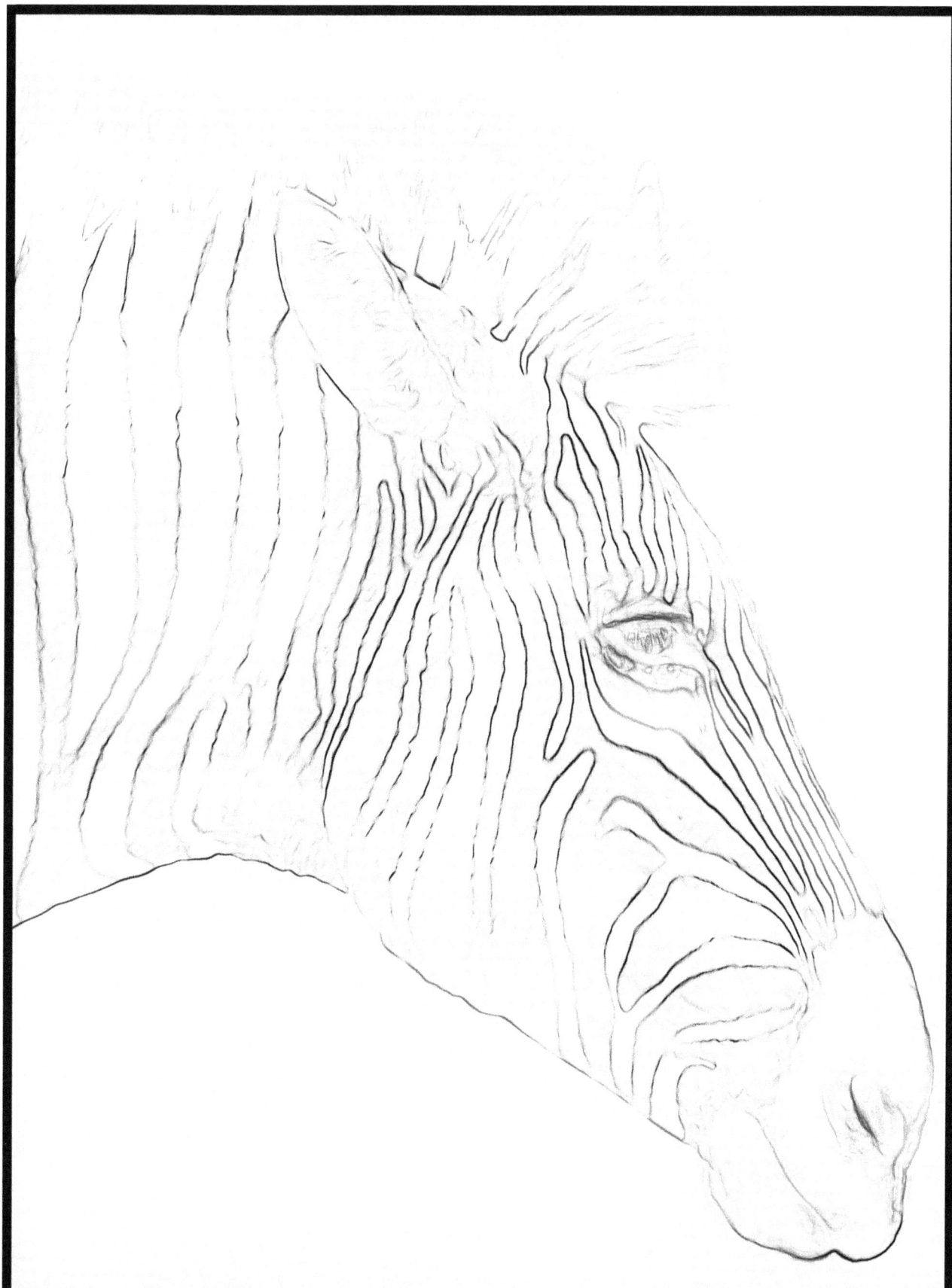

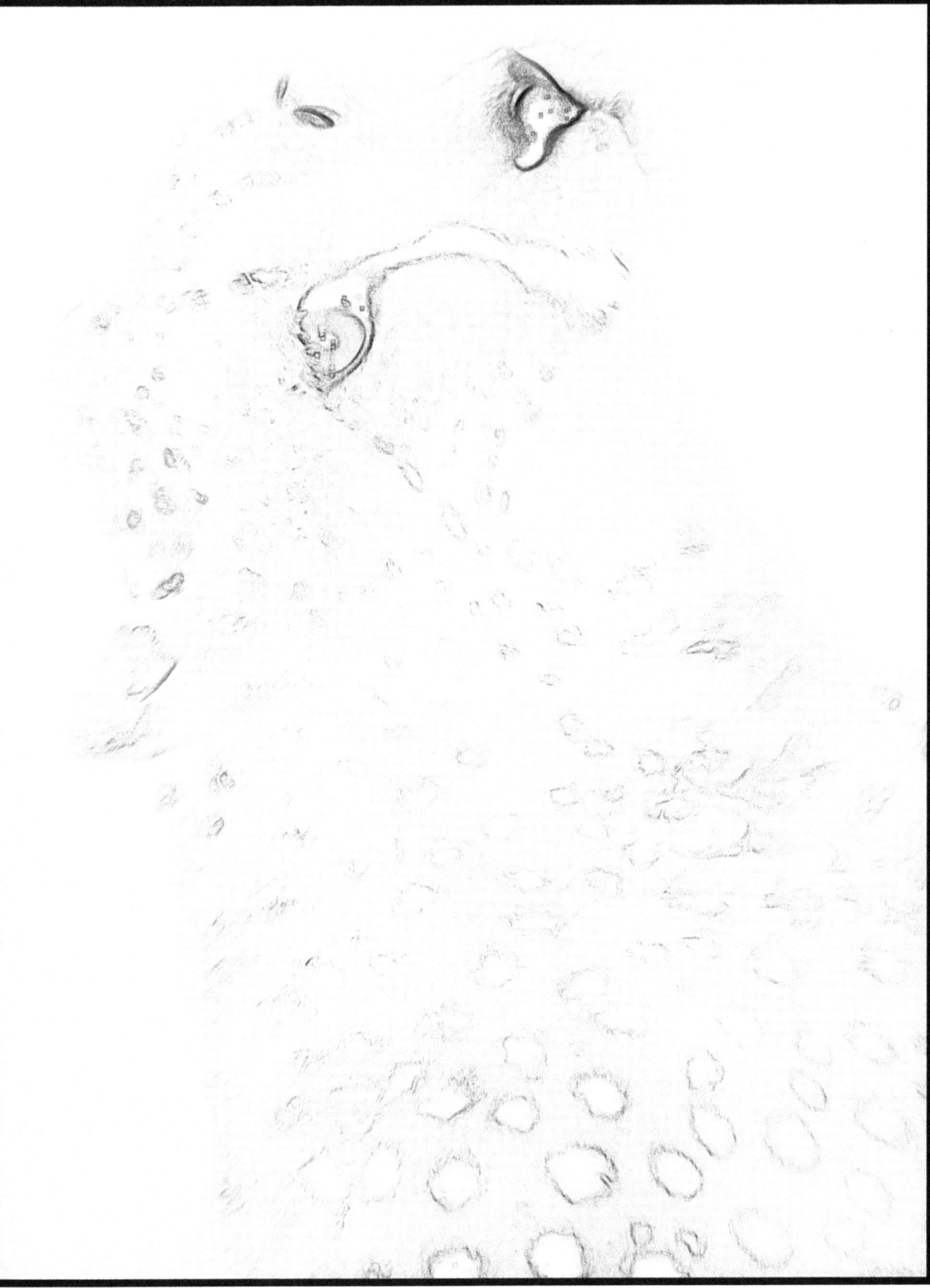

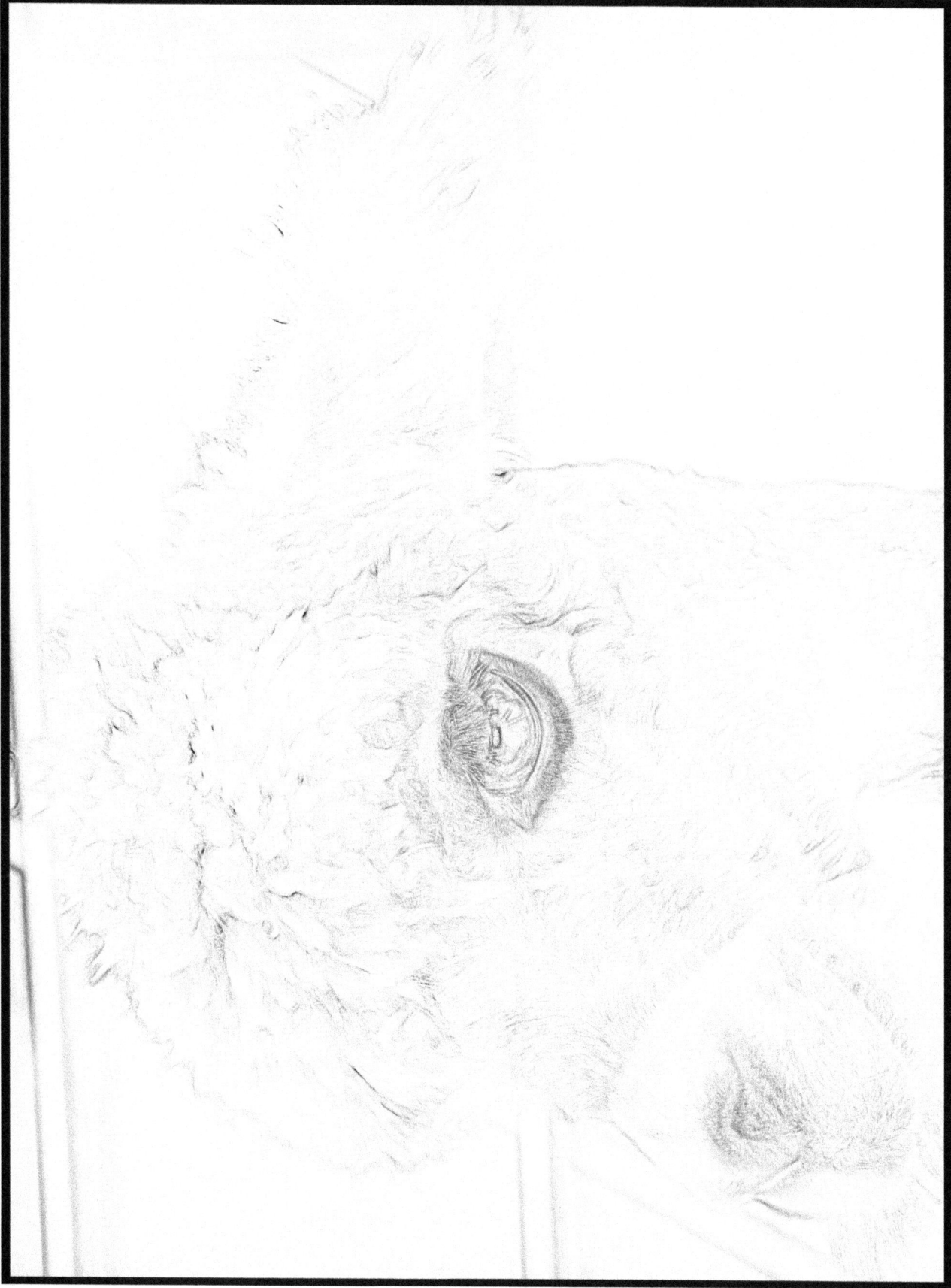

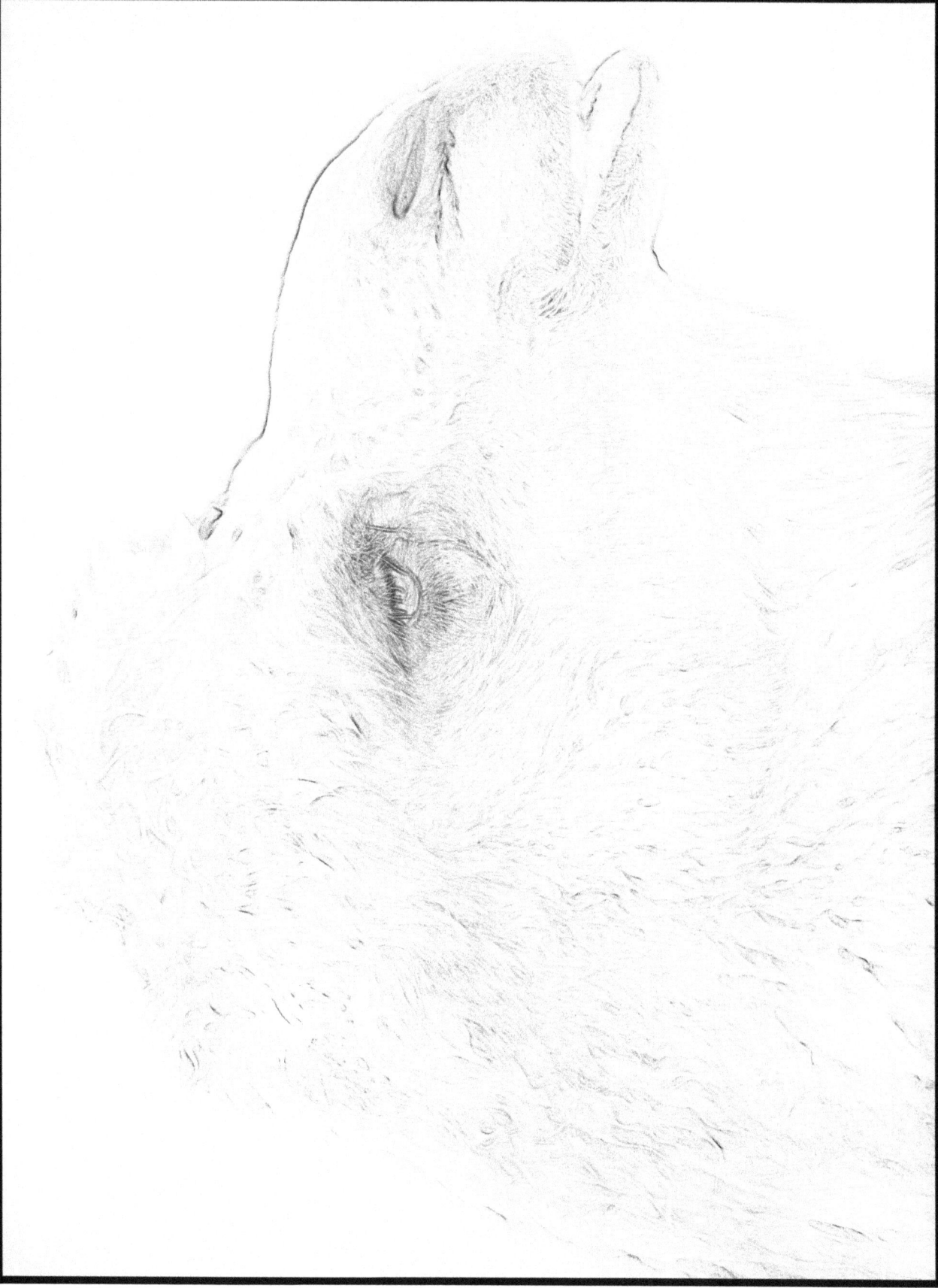

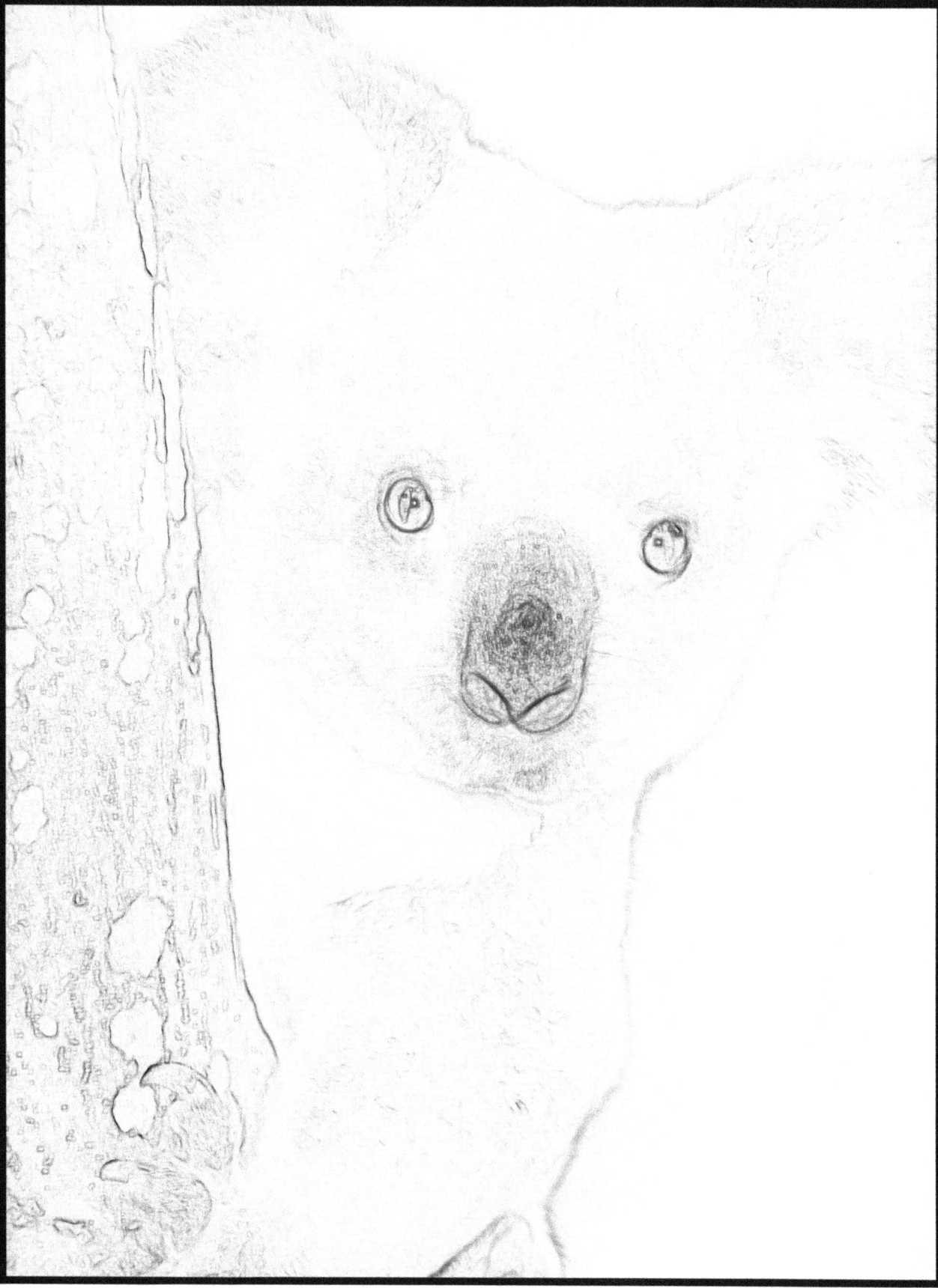

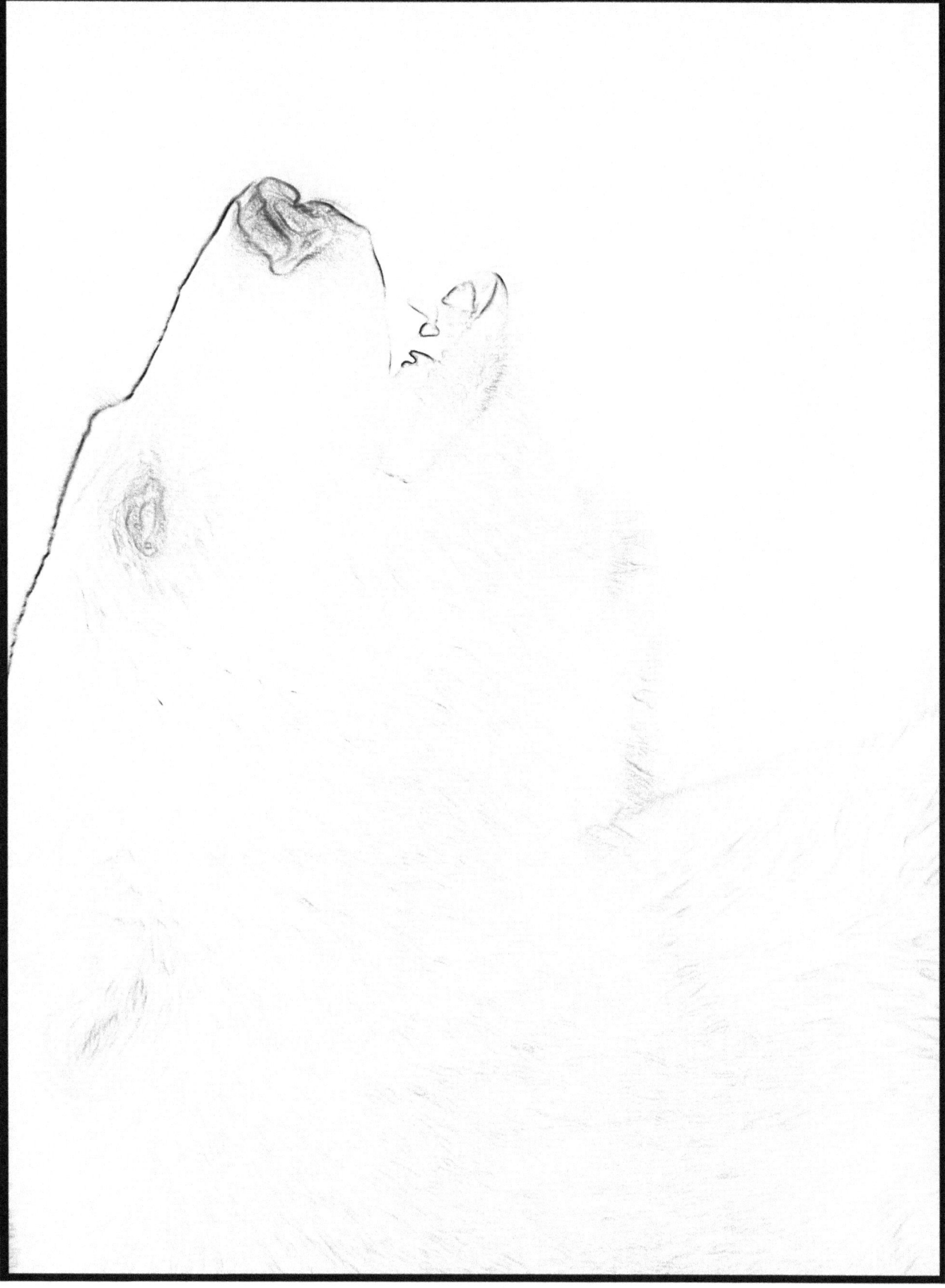

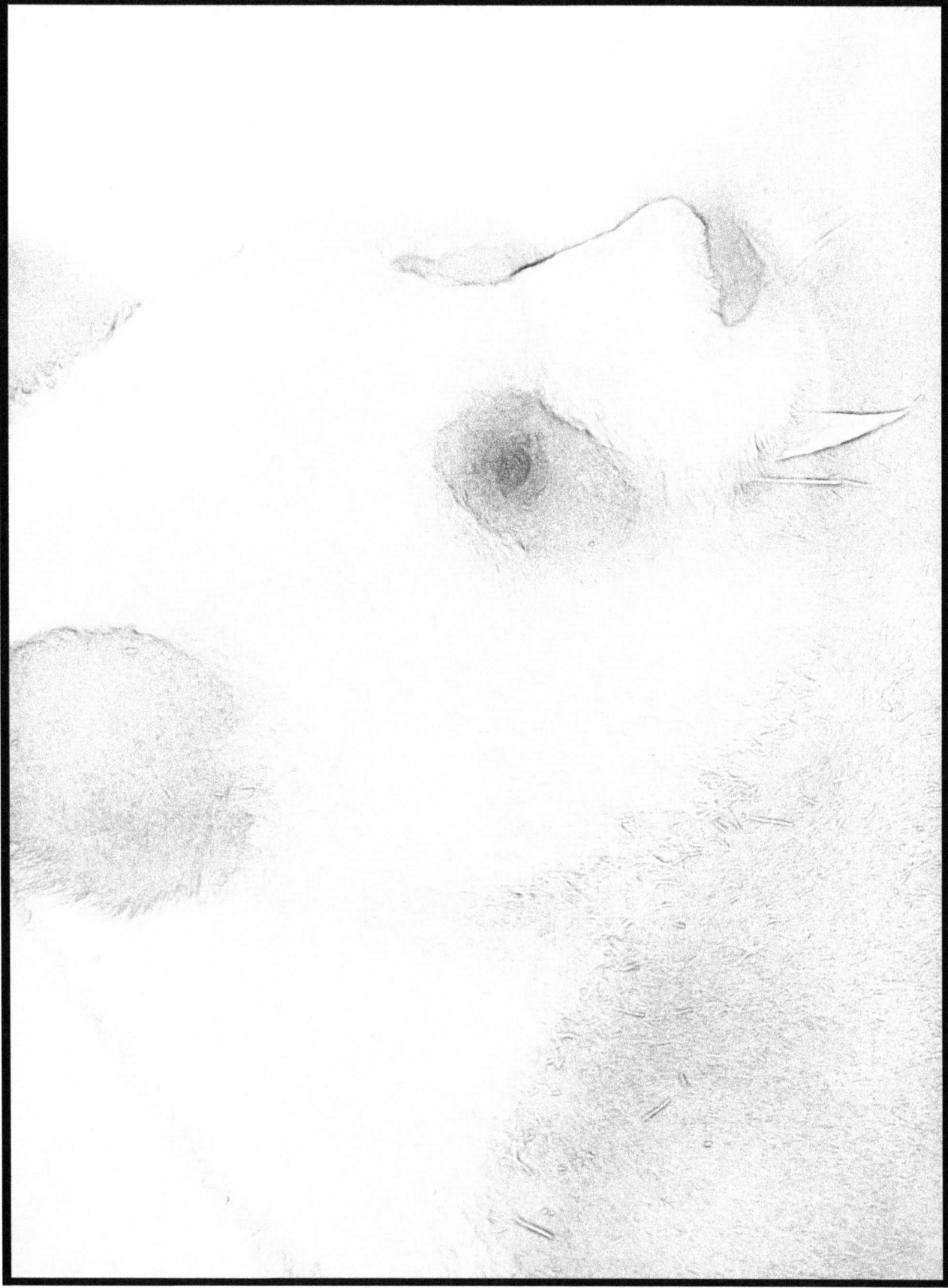

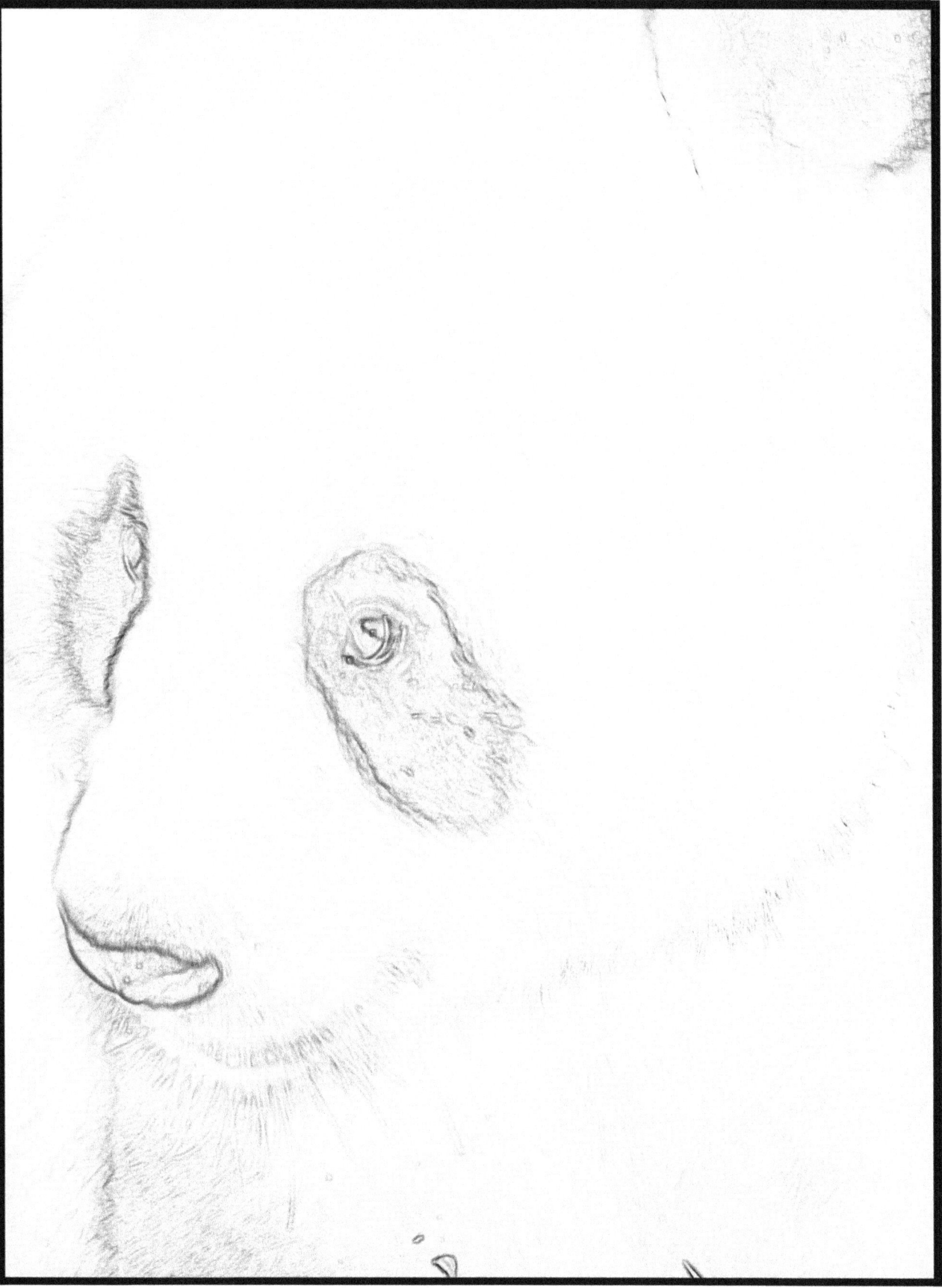

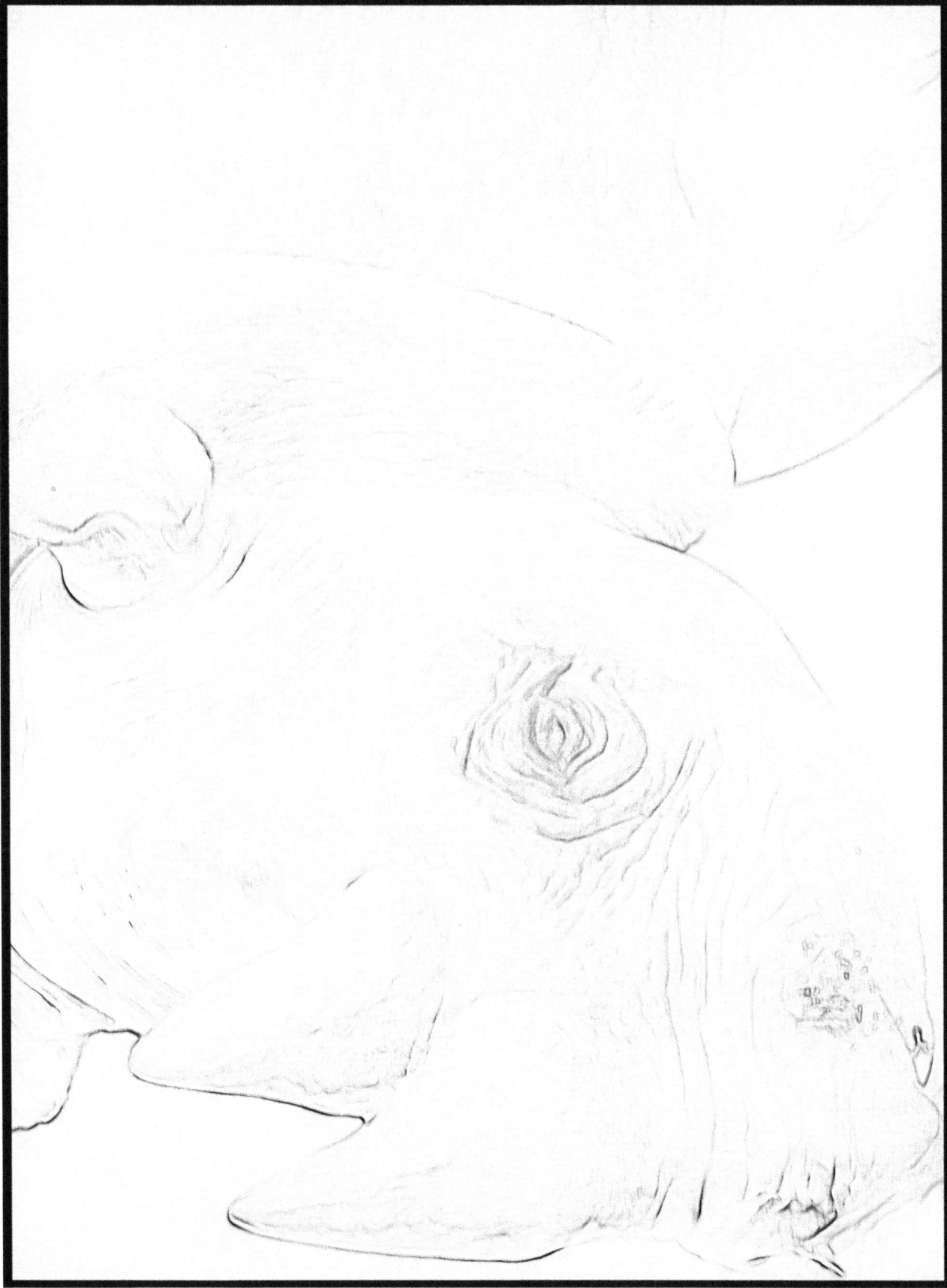

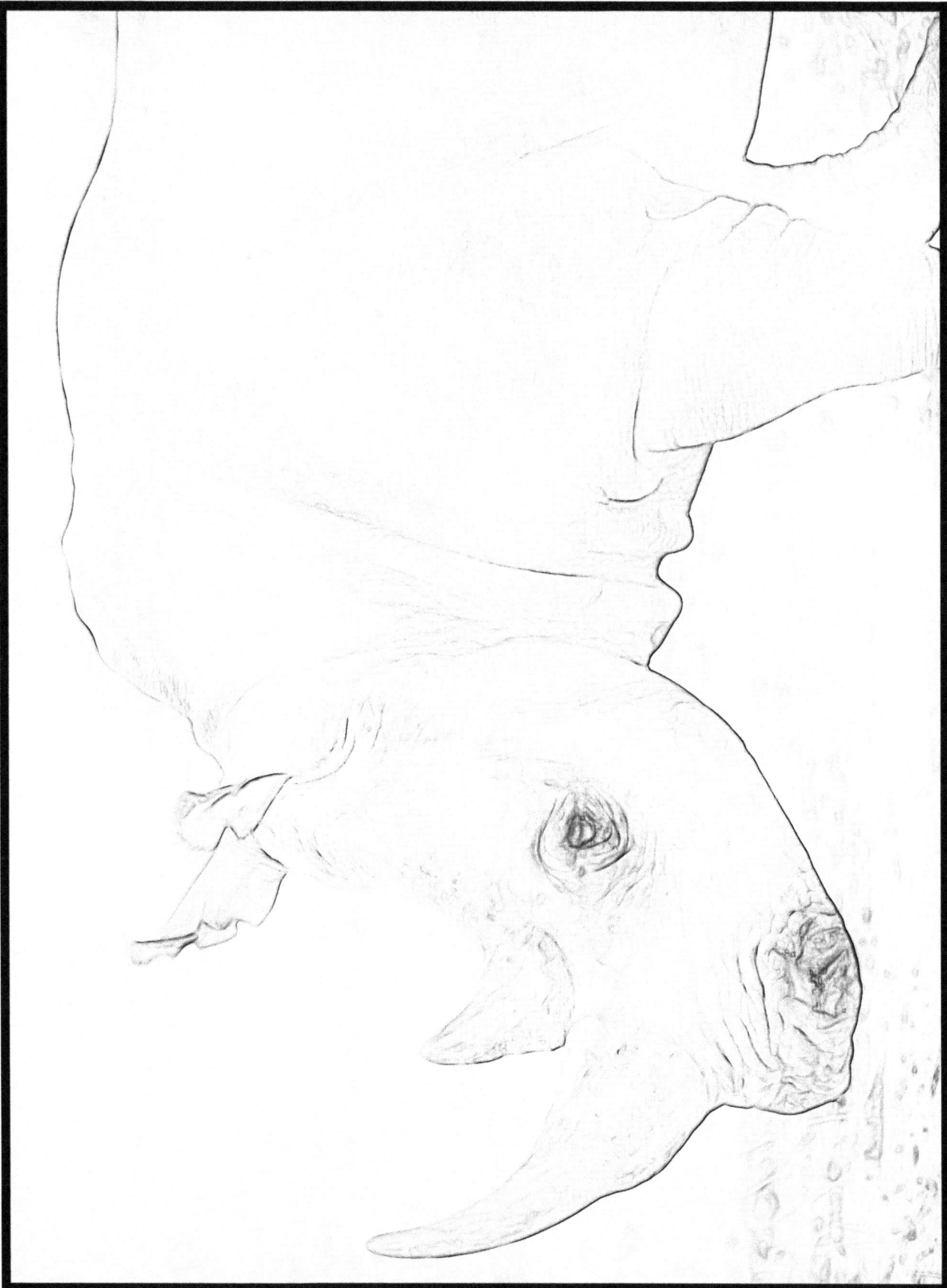

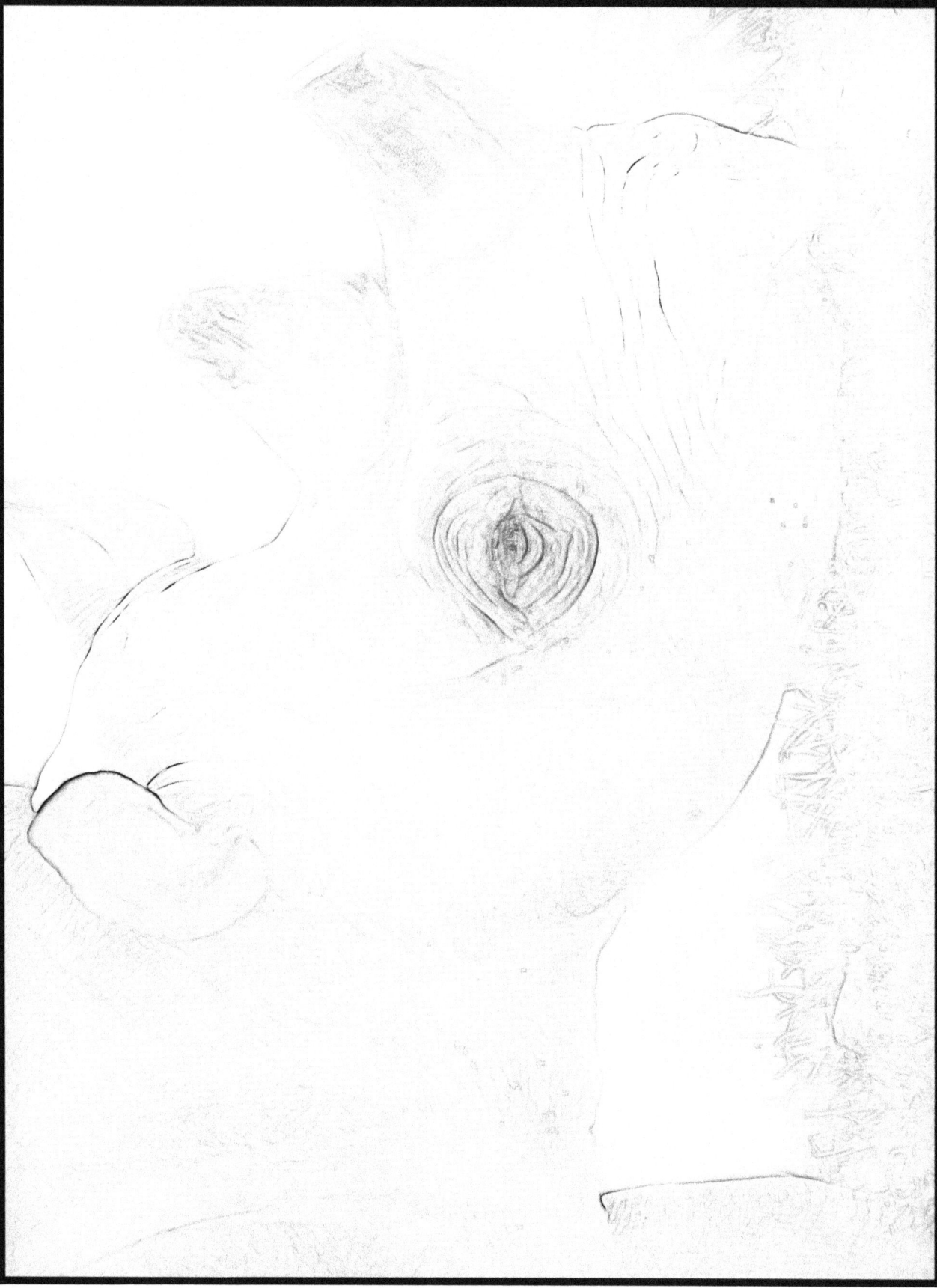

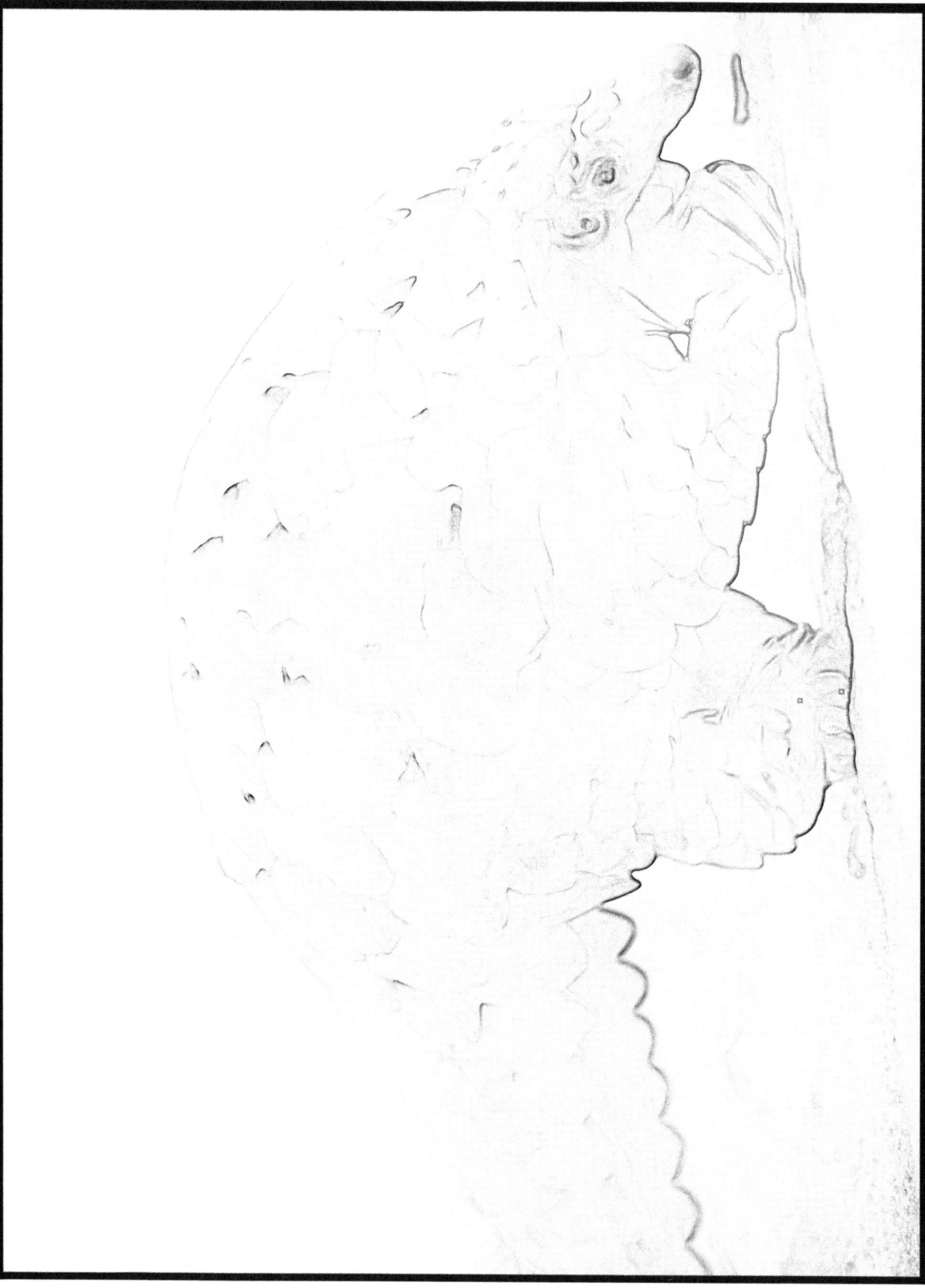

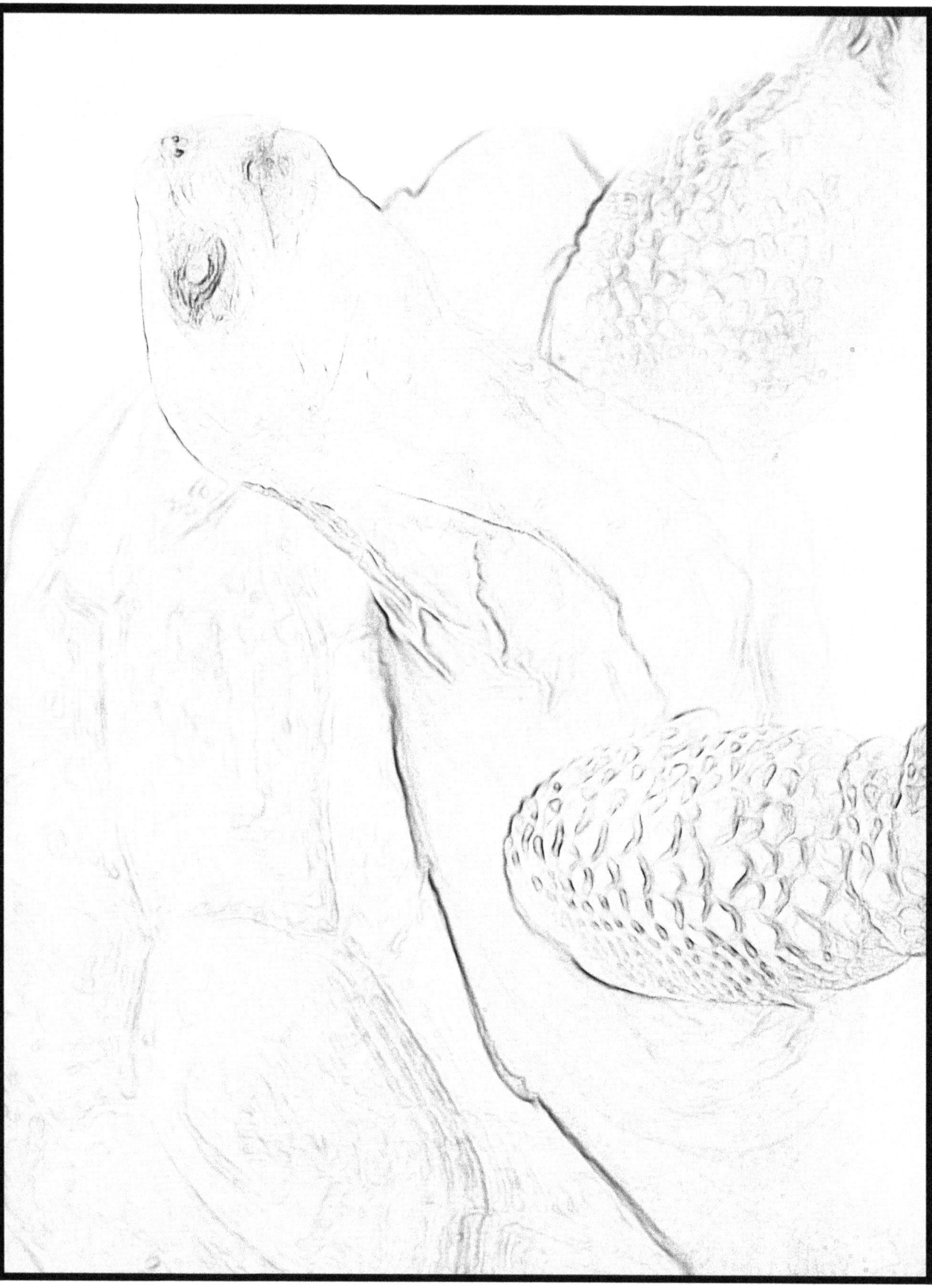

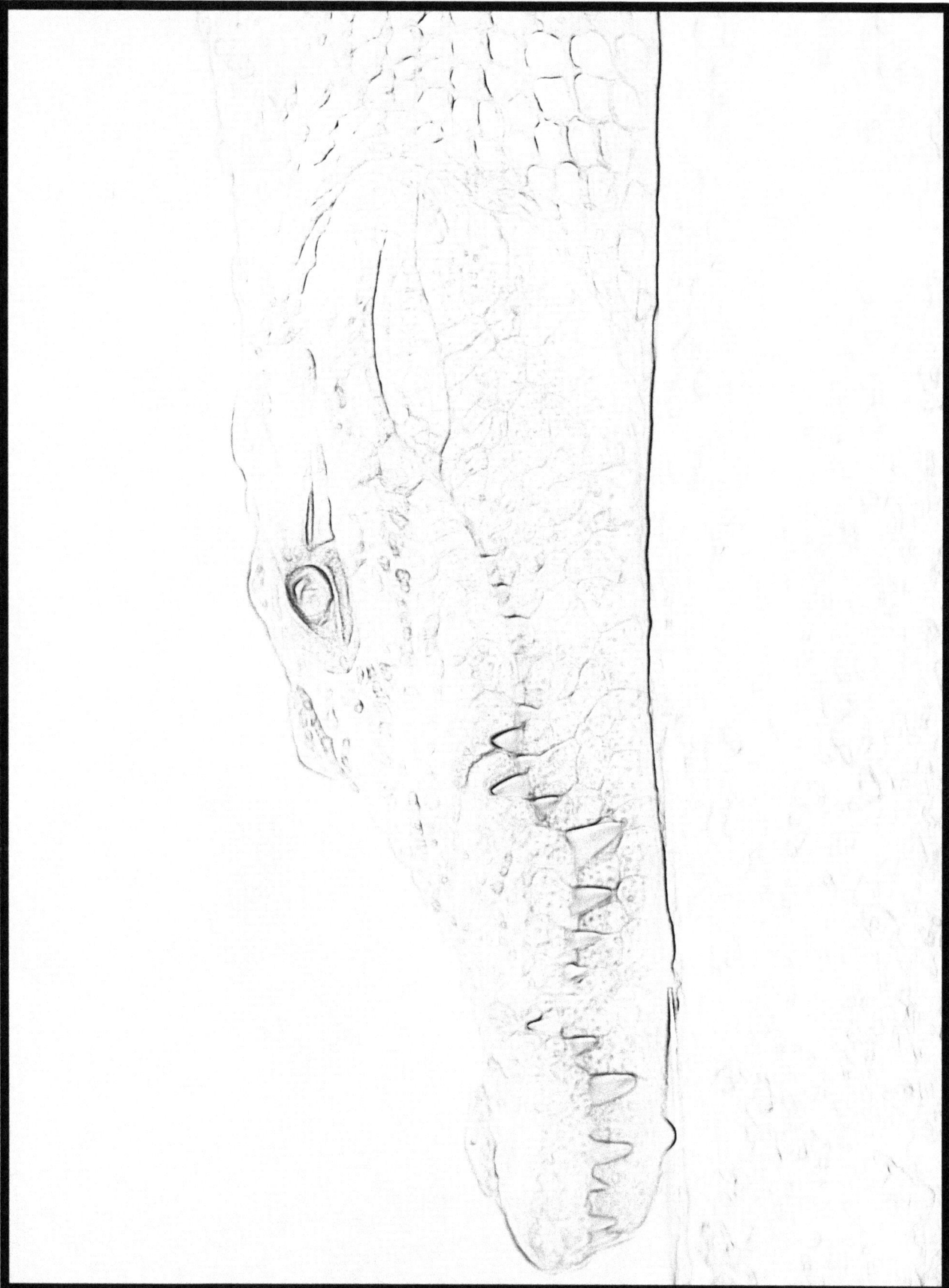

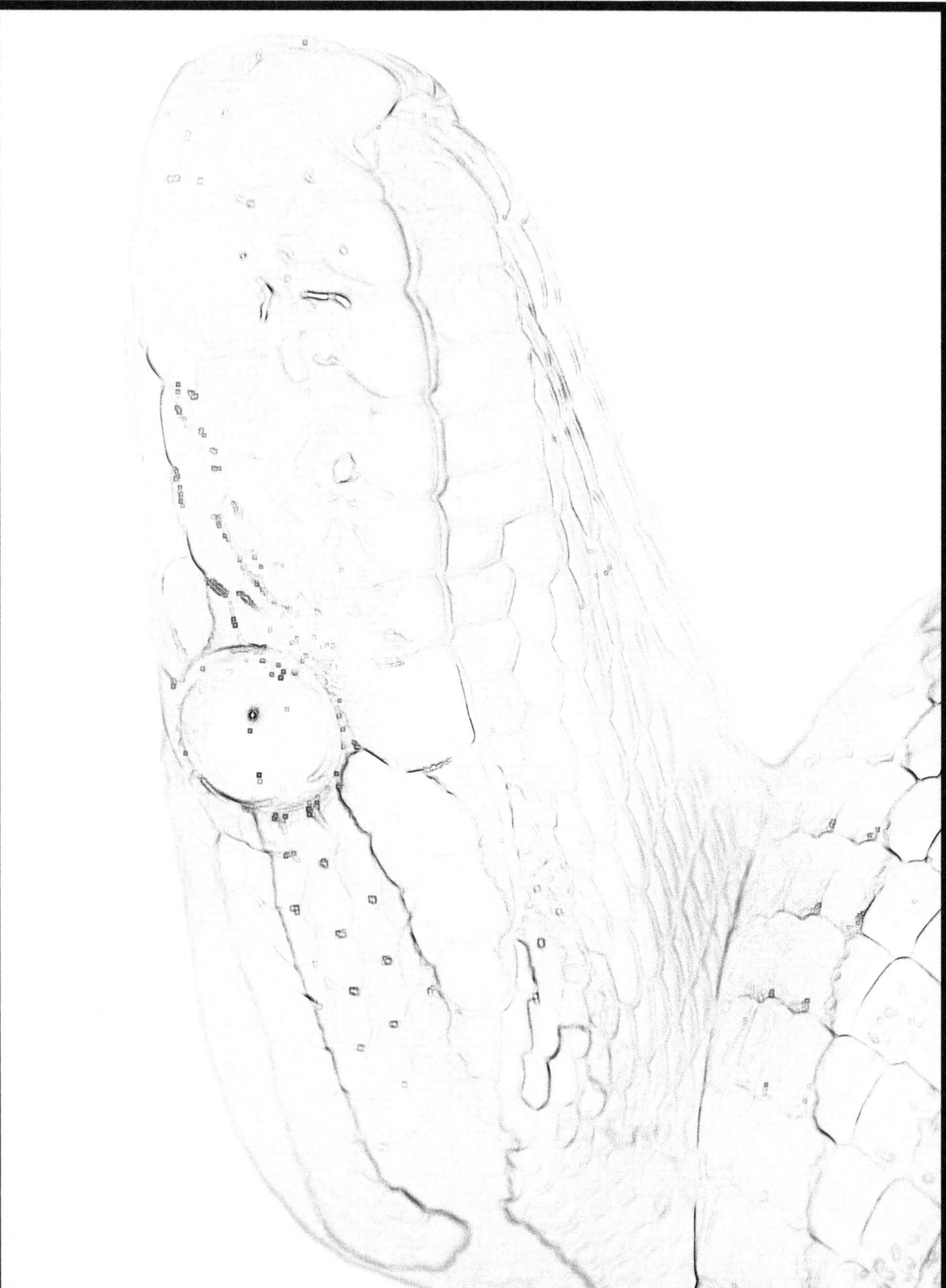